VOLCANOES

To Kīlauea, Haleakala, Stromboli, Arenal, and Vulcano — the volcanoes I have visited and been inspired by — and to all the scientists who have risked (and lost) their lives studying volcanoes to help protect the rest of us. —NCB

Dedicated to my wife, Jaclyn; my dog, Toby Tobsters; and all my Linda City Teammates. —KC

Acknowledgments: The author would like to thank Dr. Emily M. Klein, professor of Earth and climate sciences at Duke University, Bruce Nelson, professor of Earth and space sciences at the University of Washington, and Dr. Cindy Lee Van Dover, professor of marine science and conservation at Duke University, for their expert manuscript development and review.

Photos ©: 38 top: Jeremie Richard/AFP via Getty Images; 38 bottom: Zulkarnain Ginting/Anadolu Agency/Getty Images; 39 top left: Jim Sugar/Getty Images; 39 top right: Ulet Ifansasti/Getty Images; 39 center left: Erwan Gardan/500px/Getty Images; 39 center right: R. Lamb/ClassicStock/Getty Images; 39 bottom left: Claudio Cruz/AFP via Getty Images.

Library of Congress Cataloging-in-Publication Data available

ISBN 978-1-338-87467-9

10 9 8 7 6 5 4 3 2 1   24 25 26 27 28

Printed in China   38
First edition, August 2024

Book design by Sarah Dvojack
Art direction by Brian LaRossa

The text type was set in Americana and Futura. The display type was set in Americana.

VOLCANOES

WRITTEN BY
Nell Cross Beckerman

ILLUSTRATED BY
Kalen Chock

Orchard Books ◆ **New York**

Plates shift.

Land tilts.

Gas seeps.

What

is

coming?

A rumble.

A tremble.

A grumble.

Growing,

growling,

getting hot.

When will it . . .

A volcano is a vent in the crust of Earth that releases lava, steam, ash, and rocks. While volcanoes exist all over the world, most are dormant or extinct, meaning they are not actively erupting, or very likely will never erupt again. But there are plenty of active ones — at least 20 volcanoes could be erupting right this very second.

Oozing

like

thick,

hot

honey.

Glowing taffy.

Surface cooling to

hard

black

while red flows

underneath.

Don't
step
in
the

crack!

Glowing red lava can reach more than 2,000 degrees Fahrenheit (1,093 degrees Celsius). Lava leaves volcanoes in different ways. Gently flowing lava is called by the Hawaiian name pahoehoe. Other eruptions are more explosive, ejecting large chunks of lava called bombs. Because of their unpredictability, lava bombs are dangerous — in 2018, a lava bomb the size of a bowling ball set a man's porch on fire.

Collecting

sweltering

samples.

Hiking in

shoe-melting

temperatures.

Risking it all

to learn more.

Superheroes?

Scientists.

Scientists who study volcanoes are called volcanologists. They try to predict when volcanoes might erupt. By measuring tiny earthquakes and subtle changes in a volcano's shape, they can figure out what kind of "mood" a volcano might be in. To protect themselves near lava, they wear special suits and boots to keep out the incredible heat. Safety is key.

Fountains of fire

and rivers of lava

erupt

and interrupt

life.

Forming islands.

Creating craters.

Ever-changing.

Our planet is

dynamic.

When lava cools, it becomes hard rock. Over time, multiple eruptions form new areas of land. For volcanoes in the middle of the ocean, this land creates an island. Millions of years ago, a string of volcanoes began erupting in the Pacific Ocean, forming what we now call the Hawaiian Islands.

One island

disappeared overnight

with a

deafening detonation,

bursting a bevy of

eardrums.

But it wasn't the

ash,

gas,

lava,

or burning

rock rain

that wiped

everyone

out.

The real danger was . . .

TSUNAMIS!

On August 27, 1883, the volcanic eruption of Krakatoa in Indonesia created the loudest sound in recorded human history. The sound wave generated by the explosion circled the whole world multiple times. On the *Norham Castle*, a ship floating 40 miles (64 kilometers) from the blast, over half the crew's eardrums burst! The eruption also created tsunamis—giant waves that, in this case, were over 120 feet (37 meters) high. These tsunamis wiped out surrounding populations. One wave brought the population of Merak, a nearby port town, from 2,700 to two. In total, 36,000 people died from the tsunamis caused by the Krakatoa eruption.

Plates bumping,

bubbling,

and burning at the

edges.

452

spots of hot,

arcing through the ocean.

Can you see it?

The Ring of Fire.

The rocky outer layer of Earth is made up of huge pieces that fit together like a gigantic jigsaw puzzle. We call these pieces tectonic plates, and they slowly move over the surface of Earth. One of these plates, the Pacific Plate, is surrounded by most of the world's active volcanoes and is called the Ring of Fire.

Scientists thought they knew why

hotter water

came from the black,

cold

bottom

of the ocean.

But they didn't know they would find

a whole new world.

While volcanoes can be destructive, mid-ocean ridge volcanoes rising from the ocean floor bring life. After recording unexpected areas of warmer water temperatures in the ocean, scientists used the deep-sea submersible *Alvin* to find answers. They discovered underwater volcanoes that expelled hot, mineral-rich water through cracks and vents. What a surprise to learn that this allowed a whole community of life to exist in the darkest depths of the ocean — completely without sunlight! Now some scientists think life itself may have started around underwater volcanoes.

Uncorked from the ocean floor

an avalanche of ash,

a plume of doom,

a cinder shotput

seen from space.

In January 2022, the Hunga Tonga–Hunga Ha'apai underwater volcano in the Tonga Islands exploded with such force that the ash cloud stretched 36 miles (58 kilometers) skyward. That's about five times higher than a plane flies! The eruption caused a tsunami warning across the entire Pacific Ocean. It was the biggest volcanic eruption in more than 30 years, since the 1991 explosion of Mount Pinatubo in the Philippines. According to NASA, the explosion was bigger than that of an atomic bomb.

Look up!

What do you see?

Night sky,

filled with

stars,

meteors,

the moon,

planets,

and . . .

Extraterrestrial volcanoes.

The largest known volcano is Olympus Mons, on Mars. At 374 miles (602 kilometers) wide and 16 miles (26 kilometers) high, it is 100 times larger in volume than Earth's largest active volcano, Mauna Loa. Our solar system is home to a number of volcanoes. Not all of them erupt lava: In fact, the volcanoes on Saturn's moon, Enceladus, are cryovolcanoes that shoot ice particles, water vapor, and other materials into space.

Volcanoes

everywhere.

Space.

Land.

Ocean.

Do these megaphones of

magma

hold the answers to our

biggest questions?

How was Earth created?

How are planets formed?

How does life start?

What is coming?

Maybe you
will find out.

AUTHOR'S NOTE

Have you ever pretended that the floor is lava, where you jump from the couch to a chair so you don't fall in the lava pit? I loved that game when I was a kid. It became my dream to see real lava.

In 2003, I finally got the chance. I visited the Kīlauea volcano on Hawaii's Big Island, and I got to see real red-hot lava! Following safety guidelines, we started out, picking our way over sharp, cooled lava, trying to get closer to where the lava river was streaming over a cliff into the ocean. Although the day was ending and the sky was getting darker, I realized I was feeling hotter.

I looked down. In between the black, cooled cracks in the ground, I saw red glowing up at me.

I froze.

I couldn't believe it. THE FLOOR WAS LAVA! While I was safe where I stood, I knew if I poked my finger down into the crack it would melt right off! Once I got my voice back, I screamed, "Lava! Lava!" and asked a friend to take a photo of me. It was so thrilling. The night turned black. The stars came out. The waves rumbled in the distance. The lava glowed brighter. I felt like I was on the edge of the earth, farther away from civilization than I had ever been.

I kept this powerful, magical memory close to my heart for twenty years, when it inspired me to write this book. Hopefully, you have enjoyed reading and learning about volcanoes and feel inspired to go on your own nature adventures and explore the world.

You can do it!

Nell at the Kīlauea volcano in Hawaii

ILLUSTRATOR'S NOTE

All the amazing efforts that led to the creation of this book only further my appreciation for our planet and all the things that inhabit it.

I believe it's important we continue to find ways to take care of our planet because it's the only one we have. I hope this book will inspire you to be curious about the world around you.

Further Reading

◆ Batten, Mary. *Life in Hot Water.* Peachtree Publishing, 2021.

◆ Benediktsson, *Ævar Þór. Stranded!: A Mostly True Story from Iceland.* Barefoot Books, 2023.

◆ Cerullo, Mary M. *Volcano: Where Fire and Water Meet.* Capstone Editions, 2021.

◆ Ford, Robert J. *Volcano: Live, Dormant and Extinct Volcanoes Around the World.* Amber Books, 2021.

◆ Gibbons, Gail. *Volcanoes.* Holiday House, 2023.

◆ Kulekjian, Jessica. *Kaboom! A Volcano Erupts.* Kids Can Press, 2023.

The Different Types of Volcanic Eruptions

In much the same way as there are different types of storms, there are also different types of volcanic eruptions. Before we look at these eruptions, we need to understand the two basic types of volcanoes — **red** and **gray**.

Red volcanoes may sound like the more dangerous of the two, but they are actually less destructive. Although red volcanoes eject lava that can destroy buildings, the lava flows out at a relatively slow speed.

Red volcano

Gray volcanoes are much scarier. The thick magma inside a gray volcano is full of gas, causing pressure to build inside the volcano until it explodes. It shoots a huge gray cloud full of ash, gas, and rock high into the air. This cloud then falls back to Earth, forming what's known as a pyroclastic flow. These flows are very hot — over 1,000 degrees Fahrenheit

Gray volcano

(538 degrees Celsius) — and can travel across the ground at more than 100 miles per hour (160 km per hr). In 1902, Mount Pelée exploded in Martinique, creating a pyroclastic flow that killed nearly 30,000 local people.

The power of an eruption, particularly when it comes to gray volcanoes, is measured by the Volcanic Explosivity Index (VEI). Some examples are:

Hawaiian Eruptions (VEI: 0–1)

As the name suggests, these are the kinds of volcanic eruptions you'll find in Hawaii. Although these are nonexplosive, or only weakly so, they create red-hot lava rivers.

Strombolian Eruptions (VEI: 1–2)

Named after Stromboli, an active volcano near Sicily, Italy, these eruptions hurl lava tens or hundreds of yards into the air, making glowing arcs.

Vulcanian Eruptions (VEI: 3–4)

These are much more explosive eruptions that can throw ash and rock high into the air, forming a massive cauliflower- or mushroom-shaped cloud. In November 2019, the Mexican volcano Popocatépetl experienced a vulcanian eruption that forced a passenger flight from Amsterdam to turn around and head back home.

Pelean Eruptions (VEI: 5)

In this type of highly explosive eruption, a lava dome forms and devastating pyroclastic flows course down the slopes. Glowing avalanches of gas and ash as hot as 1,300 °F (700 °C) burn everything in their paths. The successful prediction of the 2021 eruption of La Soufrière in the Caribbean resulted in 16,000 people being safely evacuated.

Plinian Eruptions (VEI: 5–8)

These eruptions are truly destructive, shooting red-hot ash clouds up to 34 miles (55 kilometers) into the air. During the 1980 eruption of Mount St. Helens in Washington State, the volcano shot ash so high that winds carried it all around the world!

The Big Questions That Volcanologists Are Trying to Answer

Can we predict a volcanic eruption?

Although volcanologists can study a volcano and give a rough guess as to when it might explode, these guesses aren't precise at all. With so many people living near active volcanoes, we need to be able to predict eruptions more accurately.

How quickly can a volcano "reload"?

If a volcano has erupted, how long will it take for that volcano to store up enough magma to erupt again? It might be months, it might be hundreds of years — we just don't know.

How can we be better prepared for volcanic eruptions?

Besides predicting when volcanoes will erupt, scientists and government officials are studying methods to evacuate people quickly and restore essential services like electricity, heating, food, and drinking water after disaster strikes.

Even More Facts About Volcanoes!

- More than 800 million people live in areas that could be damaged by volcanic eruptions, and this number is growing.

- The ash from volcanoes is excellent at producing nutrient-rich farmland — which is one reason why so many people live near volcanoes in the first place.

- There are about 1,500 active volcanoes, not counting those under the oceans. Most of them are on the Ring of Fire in the Pacific Ocean. Dozens erupt every year.

- The word *volcano* comes from the name of the Roman god of fire, Vulcan.

- The deadliest eruption ever was Mount Tambora in 1815, which killed 10,000 people in Indonesia. The volcanic ash killed crops, leading to more deaths. Ash and sulfur traveled the globe, causing crops to fail in many other countries. Including the people who died in the aftermath, the death toll reached over a million.